The Theory of
Schizophrenic Negativism

By Eugen Bleuler
Professor of Psychiatry, University of Zurich; Director of Burghölzli Asylum

Translated by William A. White, M.D.
Superintendent of the Government Hospital for the Insane, Washington D. C.

Logo art adapted from work by Bernard Gagnon

ISBN-13: 978-0-359-74911-9

First published in 1912

Contents

The Theory of Schizophrenic Negativism

CONTENTS

The theories of negativism that have been advanced heretofore are incorrect or unsatisfying. Negativism is a complicated symptom, having in some cases, many cooperating causes.

The predisposing causes of negativistic phenomena are:

1. *Ambitendency,* which sets free with every tendency a counter tendency.

2. *Ambivalency,* which gives to the same idea two contrary feeling tones and invests the same thought simultaneously with both a positive and a negative character.

3. *The schizophrenic splitting of the psyche,* which hinders the proper balancing of the opposing and cooperating psychisms, with the result that the most inappropriate impulse can be transferred into action just as well as the right impulse and that in addition to the right thought, or instead of it, its negative can be thought.

4. *The lack of clearness and imperfect logic of the schizophrenic thoughts* in general which makes a theoretical and practical adaptation to reality difficult or impossible.

On the ground of this disposition there may occur direct negativistic phenomena in such a manner that positive and negative psychisms replace one another indiscriminately, only the incorrect reactions standing out as pathological negativism.

As a rule, however, the negativistic reaction does not appear merely as accidental, but as actually preferred to the correct reaction.

In ordinary *external negativism* which consists in the negation of external influences (Ex, Command) and of what one would normally expect the patient to do (Ex. Defecation in the closet instead of the bed), the following causes are at work:

(*a*) The autistic withdrawing of the patient into his phantasies, which makes every influence acting from without comparatively an intolerable interruption. This appears to be the most important factor. In severe cases it alone is sufficient to produce negativism.

(*b*) The existence of a hurt (negative complex, unfulfilled wish) which must be protected from contacts.

(*c*) The misunderstanding of the surroundings and their purpose.

(*d*) Direct hostile relations to the surroundings.

(*e*) The pathological irritability of the schizophrenic.

(*f*) The pressure of thought and other difficulties of action and of thought, through which every reaction becomes painful.

(*g*) The sexuality with its ambivalent feeling tones is also often one of the roots of negativistic reaction.

Inner negativism (contrary tendency opposed to the will, and intellectually opposed to the right thoughts) is accounted for, in large part, by ambitendency and ambivalency, which in view of the inner splitting of the thought renders intelligible a slight preference for the negativistic reaction. Very pronounced phenomena of inner negativism probably have other cooperating causes, which we, at the present time, do not know.

A conclusive explanation of all negativistic phenomena would be premature. It seems to me, however, that the falsity or unsatisfactory nature of the theories hitherto erected might be demonstrated. It is always possible to discover roots of negativism in other directions and to understand genetically, at least, a part of the symptoms grouped together under this name. A better attitude is gained in this way for further progress.

At this point we must first make clear that negativism is not a unitary symptom. The chief and predominating group is characterized throughout by the fact that the patient, by outside influences, by command, will not do precisely, what under normal conditions would be expected (passive negativism); or, that he does exactly the opposite (active negativism). A command is not executed, most often after a clearly repulsing mimik. If one tries to bring about a desired movement passively (raise the arm, sit up to slip on clothes) they show opposition, seek to get away, resist often with abuse and blows. The patients will not stand up, will not go to bed, if it is desired of them; they will not sit at the place assigned to them, will not eat the food offered them, they take the soup with the spoon for the preserves, and the preserves with the soup spoon; they satisfy natural needs out of time and place. From simple opposition to the active execution of the opposite of what is expected there are all gradations.

Not even this circumscribed group gives an impression of unity. Most patients indeed combine their negativistic actions with an affect of irritability, vexation, anger. This emotional reaction is, however, not a necessary component. If the negativistic action is simply the contradiction of a custom, if it is not interfered with from without, the previous mood is usually maintained; the patient lies down, with apparent indifference, in the bed of his neighbor; in some cases one sees even a certain mirth over a successful trick. Repression first awakes irritability in these cases.

Often the patients maintain their indifference in spite of opposition; it may be that very strongly negativistic patients are permanently euphoric and do not come out of this mood, while they resist with bites, scratches, and blows the invitation to shake hands; their defense is sport for them like a jolly play. More commonly the whole behavior looks like that of a flirt; women patients watch the physician, as if they were waiting for him to offer them his hand, or bring forward a request, so that he must busy himself with them, and then, in their negation, behave like a maiden who stimulates her lover, but tries to appear as if she were keeping him off. At other times the negativism has a plainly erotic character, sometimes in the agreeable sense of a love-play, sometimes in an unpleasant sense, as the aversion to an attack, and often in both directions at once.

Besides this outer negativism there is also an inner, which most frequently affects the will. The patient can not do exactly what he wishes to do. In the stage between thought and expression an inhibition, a contrary impulse, or a cross impulse can make the action impossible. So we see patients who rush to take a proffered bit of food, stop half way between plate and mouth, and finally refuse the morsel; with every other act the same results follow. If they start to shake hands: at any point the action may not only stop but the hand, as the result of a contrary impulse, may be placed behind the back.

Often the patients frustrate the results of an act by other movements. They stretch the arm out in order to proffer the hand but flex the forearm and hand so that the hand can not be taken; or on the request to show the tongue they put it out but turn away the head. In some of these cases of simultaneous obedience and disobedience one usually sees the external negativism. But undoubtedly the phenomenon occurs as a pure will-negativism. I have noticed it when the patient spontaneously occupies himself in something without outside invitation, for example in eating; mostly I have observed it in piano playing. They reach out for the stroke and strike down with the forearm, but towards the end of the movement dorsally flex the wrist to the maximum, so that the fingers do not reach the keys, or the patient turns the eyes to one side, in order to observe something, and at the same time turns the head to the other side (or the reverse).

Cross impulses assert themselves, in that, instead of a willed or begun act another is carried out; the patient starts to take a spoon (to eat), the begun movement is changed however and he takes the fork, puts it in the bread basket or does something else equally peculiar. These cases present all transitions to the apraxoid appearances of schizophrenia, which on their side again have different roots.

Not infrequently negativism shows itself towards a task which has already been completed. The patients destroy what they have made. Sometimes as if in anger, sometimes as if from a free resolution, sometimes compulsively, resenting it in the doing.

It is very difficult to get a clear idea of the subjective process of this will-negativism. Very few patients offer any explanation. It is certain, nevertheless, that some are aware of the disturbance, but not others, and that all possibilities actually occur with regard to the psychological point where this sets in. The patients suddenly no longer will what they have just intended, or they suddenly will the opposite; their motive may come into consciousness or not; the goal idea becomes altered. This can, however, also remain the same, while the centrifugal impulse becomes disturbed somewhere in the tract z-m, which one can not conceive sufficiently long and complicated (compare for example Liepmann's researches on apraxia). Here the patients of course become more or less aware of the disturbance; but some bear it with the thoughtless in-difference of schizophrenia, others feel it as a peculiarity which has befallen them, and conceive it, sometimes as something abnormal, sometimes as an influence from outside. Not at all seldom the negativistic impulse is transferred into hallucinations, which then, like other sensory falsifications, are interpreted subjectively in the most varied manner. A catatonic, for example, who will say something, hears his neighbor command, "Hold your mouth." Another, who will eat, the voices forbid, or say, it would not be right; if he does not eat, it is again not right; he asks despairingly, what in heaven's name he may do. To a third the voices always say the opposite of what he must do. A fourth receives hallucinatory commands for example, to write a letter; as he is about to obey, the voic-es forbid him. He calls these hallucinations very signifi-cantly "plus and minus voices." There is probably little difference in principle when the negativism is transferred into delusions. H one requests such a patient to eat, stand up, walk, he does not do it. Afterwards he complains that he gets nothing to eat, that the physician compels him to lie in bed, forbids him to walk. The commonest negativ-istic delusion is in general that the patient believes it is forbidden him (under threats of danger or temporary or everlasting punishment) to do what he wishes. It is often shown from the change of the ideas and from the incor-rect or artificial causal connection, that the delusion is in reality secondary, springs from the negativistic attitude, and so only apparently accounts for the negativistic behavior.

Intellectual negativism, negation of thought content, is the least known of all. Naturally it can only become perceptible as opposite thoughts; it will hardly be possible to demonstrate the existence of a mere negativistic resistance against the contents of the thought. We find patients, who for each thought must think the opposite, or instead of a thought, imagine its negation or its opposite. An intelligent and philosophically accomplished catatonic said, "If one utters a thought, one sees always the opposite thought. That reinforces itself and extends so quickly that one does not know which was the first." Others complain that the thought comes to them "that is cold." when they touch something warm, and the like. One of our patients who was still able to work and was not confused had at times lost the feeling for positive and negative: she praised and found fault with her possessions, her husband, etc. in one breath, so that it was not possible to bring out what she really meant.

If a patient retracts his own declaration, at times right afterwards in an agitated, pathetic tone, one can relate it just as well to negativism of the will as to that of the intelligence; he has come to the institution to get evidence — no, he wishes no evidence; and so forth.

In intellectual negativism the subjective side of the symptoms is also very variable. Many patients experience negativistic thoughts as compulsory, others are indifferent, and again others do not notice it at all. This form of negativism is also often projected as hallucinations; the patients then often hear the opposite of what they think or what they perceive in the outer world. It may also sometimes occur that the negativistic thought at its inception is transferred into compulsive actions so that the patients must say the opposite of what in reality they think.

Occasionally intellectual negativism affects only the speech mechanism. The patients say the opposite of what they wish to say, especially they express against their will a negation, when in reality an affirmation was thought. "You are not a wretch," may be said to the physician, as a résumé of a prolonged abuse for unjust confinement. One catatonic who was told to step up on a platform in the clinic protested energetically that she would not "go down there." Patients do not by any means always notice such mistakes, not even when one tries to call them to their attention.

Probably it is a milder form of this same anomaly when the patient expresses the correct idea but in an unexpected negative form: "that is not beautiful" for "that is ugly"; "that is not ugly" for "that is beautiful." In one case, which I have been able to observe for many years, such negative expressions in the mind of the patient became a unity before which another negation could be placed. She would say, "It is not not-ugly," in order to say that something was ugly. The "not-not-ugly " was used as one expression and used with a negative to express something that was beautiful. It is conceivable, that the patient easily became confused and was no longer clear whether she affirmed or denied something; then held the listener responsible for puzzling her.

Negativism in the transference of words heard into the corresponding thoughts has not yet been observed by me. It is certainly not rare that the patients understand the opposite of what we say. That is, however, only the case when this opposite is identical with their delusions and wishes. The cases known to me are therefore ordinary examples of illusions of perception and memory.

It is a very important, and yet an often-overlooked characteristic of negativism, that it does not show itself uniformly, but at times is present and at times absent in accordance with the psychical constellation. It is quite usual that patients in their relations with other patients and with the attendants appear free from negativism, but on the contrary, they are very refractory to the physicians and their regulations. The reverse is not quite so common. To visitors also the conduct may be the contrary of the usual. Certain patients become suddenly negativistic when one touches a complex. Others, on the contrary, under the same circumstances, may lose their negativism for a time.

What we have thus far designated as negativism must appear, after the mere description, to be a symptomatological collection, made up of very different things, and after it has been pointed out that the genesis of all these phenomena is not uniform, one may ask why all this is included in one conception. Not from respect for the teachings of the past, but because we are not yet able to distinguish between the various psychic processes which call forth negativism. The most varied manifestations may be derived from the same roots, and all the varieties mentioned may occur in the same patient in such mixtures and transitions that one will never be able clearly to separate them.

It is self-evident that inner negativism can assert itself outwardly in negativistic acts. He who instead of "agreeable" thinks "disagreeable," must act wrongly, and will-negativism may lead to the same inaction or to contrary action as mere defense to outside factors. On the other hand the repelling of outer influences causes an inversion of the feeling tone, which evidences itself as inner negativism. The offering of food often causes disagreeable feelings, just because it comes from without; the declining then is obvious. But the disturbance should be sought in the negativistic vitiated emotional reaction rather than in the relation to the external world. This cannot be entirely denied because pararythmic reactions are not altogether infrequent in dementia prascox.

Negativism is thus not an elementary symptom, but a collective idea, comprising a number of symptoms, which are similar one to another, in that, in the different areas of psychic activity precisely that is left undone or the contrary is done which one would otherwise expect under the existing conditions. Negativism most commonly involves a repelling of outside influences; it can express itself, however, as an inhibition or perversion of inner processes. Not even the repelling of outside influences is always founded on the same genesis, and in a given case, we shall see, several motives operate together, in order to bring about the repulsion.

The idea of negativism is not always limited in this way. Kraepelin [2] describes it under the title of weakened influence of the will and designates it as "the instinctive resistance against every outer influence of the will." This expresses itself, according to the author, in seclusion against outer impressions, in inaccessibility to every outer communication, in resistance to every demand, which can culminate in the systematic performance of exactly opposite actions. The latter is not always simply an exaggerated opposition, a "weakened influence of the will," but probably a suggestibility in a negative sense. Kraepelin does not mention inner negativism explicitly as belonging to negativism: yet for him the blocking of the will is only a partial expression of general negativism. The Kraepelinian idea of blocking is composed of two different things. What we mean by this name is a sudden arrest of psychic events that is often observed in thinking. It is one of the usual schizophrenic symptoms and has its analogy in the arrest of thought in the healthy which is produced by some affect (terror, examination fright). Such blocking in thought and will may also occur in the absence of negativism, but negativistic disturbance of the will is conditioned by a contrary will, an "opposition." It is therefore fundamentally different although the two causes may occasionally overlap as the negativism is also colored with affect.

It appears to us that Kraepelin has laid too much weight on the seclusion from outer impressions. There are negativistic patients who are interested in everything, who tease others and generally seek stimuli from without. The schizophrenic repelling of outside influences ("Autismus," see below) does not necessarily express itself in a sensory declining, but only in the selection of the impressions and their elaboration.

Hoche [3] defines negativism' as "the systematic resistance against external influencing of the will and also against impulses arising from within." Here we must replace the "systematic" that implies too much conscious activity by "instinctive" or "impulse-like." As Kraepelin rightly says, no intellectually understood motives play a part. Further "systematic" can not indicate a continuous type of conduct for the negativism does not appear at every opportunity, and it may be added, that the resistance may lead to the doing of the opposite. Furthermore, the definition also ignores intellectual negativism.

The behavior of the psyche of the patient towards the negativistic symptoms is very variable. They may be fully united with the conscious psyche; the patients are then conscious, refuse voluntarily and then are irritable if one desires something of them, exactly like a well person, who wishes to know nothing of his environment. At the other extreme the negativistic activities emerge from the unconscious (as the voices and delusions); the patients are themselves surprised by them; they even defend themselves against them for some time; they wish to be agreeable, to follow orders, but are not able to do so. Contrary impulses and inhibitions of all sorts prevent the patients from doing what they have in view, so that commonly they believe in the influence of a strange force. All gradations intervene. So long as the patients are left alone they generally relate themselves very well to the surroundings, and may resolve that they will not now react negativistically; when the opportunity occurs, however, they are protesting and irritated; they themselves wonder at such changes of mood and affect and can not discuss the matter.

An attempt has been made to explain negativism by proceeding from the motility, from muscular disturbances. Lundborg [4] finds a similarity between the catatonic muscle phenomena and myotonia and thinks, that many patients in spite of wishing to, cannot move and therefore are apparently negativistic. He even brings the stereotypies, which lead to round-about impracticable movements, into relation with this disturbance. This parallel with myotonia, shows that the author transposes the root of negativism to a centrifugal process, and thinks of this at least as peripheral; that the departure of the motor stimulus starts from the cortex, or possibly even in the muscles. I do not want to deny, that the outer picture of negativism may be produced through not being able to act, but not through a motor hindrance, but as the result of a psychic interference, like a child, who is bashful before a stranger, from whom he can not take a bonbon, even though anxious for a sweet.

We have to do then with the inhibition of a purpose brought about by a contrary affect. This also occurs naturally in schizophrenia, but it is probably preferable not to call this phenomenon negativism in spite of the external resemblance. In spite of all my effort I have been unable to see a true motor disturbance in dementia praecox either at the root of negativism or elsewhere. At all events there is nothing to observe in many cases with negativism that one could even remotely explain as a motor symptom; for many hyperkinetic patients are negativistic and vice versa the negativistic reactions frequently lead to very energetic and active muscular movements. The hypothesis, at best, can have only limited validity.

R. Vogt [5] discusses this theory but definitely localizes the difficulty of action in the motor centers. According to him there persists (as in the catatonic brain) a tendency to perseveration which manifests itself especially in the antagonists. In this way movements are made difficult, and this condition produces in the psyche a disinclination to movements in general.

In view of the general disassociation of the schizophrenic psyche, the undoubtedly common tendency to perseveration might be assumed to affect especially the antagonists in individual cases; but no one has yet observed it. But, negativism never stands in a definite quantitative ratio to the degree of perseveration, and above all, those cases do not escape where there is no trace of perseveration, and in which the movements are in no way impeded. So Vogt's view can not be right.

Roller [6] has already expressed similar ideas to those of Lundborg and Vogt, as he likewise sought to derive the negativistic "will not" from a "can not" as the result of disturbances of innervation and besides conceived, that the contraction of the antagonists by way of their "re-innervation" called forth the will to contrary action.

Alter [7] also considers negativism a motor phenomenon. His "primary negativism" springs from schizophrenic tonic rigidity. He assumes, as fundamental, a molecular alteration in the nervous system produced by toxins which makes possible sejunction in the paths of the protagonists. The exciting cause of the sejunction is the attention, which easily inhibits what one wills. Through the sejunctive inhibition in the protagonist paths the impulse is directed to the antagonists.

The existence of a catatonic tonus, as a true motor symptom is to me very questionable. My positive observations are limited to motor phenomena elicited by psychic factors and which are again removable by psychic means. On the other hand one often feels a mild resistance in the passive movements of schizophrenics even when the patient willingly surrenders himself to all experiments. One cannot deduce negativism from this, as a strong resistance is precisely the result and not the cause of the psychic reluctance.

Active negativism can not be interpreted anyhow by a roundabout way through the antagonists. The innervation of the antagonists makes no retrogression from a progression, nor does it make a "no" from a spoken or written "yes." All of this requires quite special muscular coordinations.

The theories which explain the unwillingness to act by a motor difficulty, and which deduce, from the innervation of the antagonists, the idea and the will to an opposite action, are certainly wrong; in the first place, because a motor difficulty for the most part does not exist, and if it did, it would not be able to produce the motor phenomena of negativism; in the second place, because innervation of the antagonists only exceptionally leads to a contrary action.

Wernicke [8] considers negativism and pseudo-flexibilitas as modifications of flexibilitas cerea, "which appear with retained possibility for any voluntary influence." The attempt at passive movement is perceived within the cortex. At times it arouses the idea of the movement to be executed and renders easier the corresponding action of the will, at other times the thought of the impossibility of executing the movement arises, that is, to the idea of the movement to be executed is associated at the same time the inhibiting thought of an expected outlay of strength, which appears very great in the subjective estimation. The effect of the will thereby becomes reversed. Why at one time negative, and at another time positive ideas are awakened is not explained by this theory, just as it does not explain why a passive movement should ever arouse ideas of the impossibility of executing the movement and of the expected outlay of force, and still less, how out of it can come the exhibition of strength of an often energetic resistance. The hypothesis forgets altogether, that only a small part of the negativistic phenomena is expressed as resistance to passive movements, while expressions of protest, contrary actions, and cross impulses are much commoner. Also when one puts instead of "passive movements" "any demanded movement," this view is not rendered any more plausible. We would have to find occasionally the idea of impracticability and demand for effort at the root of negativism. This I have never found. We see on the contrary aversion to mental or physical effort quite commonly without connection with negativistic expressions; one symptom may be lacking, while the other is present, and where both are found together, one notices no parallelism in their intensity.

For the comprehension of Wernicke, his further view, accepted by others, is significant, that a partial negativism occurs in single muscle groups. Observation has never given me any proofs for such an assumption. I have learned to know negativism only as a psychic phenomenon, with its expressions governed by ideas, not by anatomical conditions. Also I have been able, up to now, to localize the motor phenomena of schizophrenia only in ideas, although obliged to assume, that one of the predisposing causes lies outside the psychic area. (Perhaps something akin to brain torpor?).

The psychic theories of negativism, for the most part, have no regard for the irregularity of its expressions. Thus the theory of Raggi [9] and Paulhan, who assume a contrast association, bringing out an action opposite to the one originally thought; or that of Sante de Sanctis, [10] in whose opinion the spirit of negation inherent in us outweighs the remnant of resistance of the ego. With such "explanations" the question is shelved behind a not very accurate circumlocution of the phenomena. Still less can we take up with the assumption of a "nolition," [11] so long as this idea is not deduced from the elementary psychic manifestations.

In France and in part in Italy, negativistic phenomena, are frequently grouped with nihilistic ideas, and explained under the name of "ideas of negation." Naturally we cannot discuss with these authors, as the two symptoms are for us totally different.

Anton [12] calls attention to the fact that many hebephrenics are pathologically suggestible and are more or less aware of it; they utilize therefore an elaborate refusal as a kind of psychic self-regulation, as a protection against unpleasant influences. For this reason negativism makes a distinction between superior persons and such as are of equal or of lower station than the patient, in that it expresses itself more fully towards the former. Negativistic behavior, apart from schizophrenia is frequently noted also by us in genetic relationship with exaggerated susceptibility partly as the second side of the same affect disposition which may express itself positively and negatively partly as an instinctive (more frequently than a conscious) protective measure. Precisely in schizophrenia, however, positive and negative suggestibility do not by any means always run parallel, one with another, in course and strength. We believe that such factors essentially cooperate in the origin of negativism, but that the symptom, however, must have still other and indeed more important roots.

Schiile [13] assumes a "contrary direction of the will," which is conditioned through anxious helplessness; it expresses itself first in simple, then in contrary (active) negativism. This "anxious helplessness" is too commonly wanting in negativistic patients for us to deduce the phenomenon from it. Yet there is something true also in this conception, in so far as lack of understanding of the environment usually leads to negativistic reactions,

Gross [14] refers first to the helplessness as causing the "affect state of negation." This alone, or in conjunction with inhibition, produces a form of negativism. According to him, there is, however, in addition a "true catatonic (psychomotor) negativism," that is, "a complex of phenomena, which form the expression of a series of psychophysical processes separated from the continuity of the ego, in no way related with the psychic processes of the conscious personality, and therefore inaccessible to any introspective after contemplation." There is thirdly a "psychic" or "total" negativism, which is compounded of the two first forms. The conception of the second form can not be correct. While it is true that schizophrenic psychisms can functionate fully dissociated from the ego-complex, this does not answer the question why precisely such phenomena become negativistic. In reality such psychic automatisms can be negativistic or not, in the same manner as conscious functioning processes. On the other hand the idea of an "affect state of negation" contains an element of truth although it is not a genetic explanation of negativism. One can ascribe to all these negation processes a common component, the negation, and the negation, as with all conflicts, is associated with an affect, so that the term cannot be entirely repudiated as a circumlocution of the affective volitionistic part of the negativism. However, the idea is not at all clear, and keeping in mind the different moods in which negativistic symptoms appear, the identity of the affective phenomena, grouped together as the "affect of negation," must be doubted.

Kleist [15] also assumes a peculiar "feeling," which he parallels with the "feelings of anxiety, of anger, of joy," thus, according to our terminology, an affect is made the foundation of the negativism. In some cases it is expressed as a painful sensation of weakness in the heart, in others as an unmotived anxiety. Why it appears, the author leaves unexplained, but that is precisely what we wish to know. I must again raise the objection that I have not been able to make out a "peculiar characteristic feeling" that was the same in each case of negativism.

On close examination the same grounds will be found as causes of pathological negativism as for the negative attitude in health. First one repulses when one does not wish to be disturbed. This is also regularly the case in schizophrenic negativism. All these patients are highly "autistic," [16] that is, turned away from reality; they have retired into a dream life, or at least the essential part of their dissociated ego lives in a world of subjective ideas and wishes, so that to them reality can bring only interruptions. Many patients state this, with full consciousness, to be the reason for their behavior. They wish to remain undisturbed within themselves, and so it is apt to vex them extremely if the attendant merely comes into the room to bring food. Their stereotypies, their peculiar attitudes and other quirks have special relation to their complexes: for them they are often the realized fulfillment of their wishes; they are not only symbols of their happiness, as one might approximately conceive from the standpoint of health, but they are the essential part of happiness itself. They have, therefore, grounds enough to defend themselves against anything likely to rob them of this treasure.

It is self-evident that autism does not express itself merely in centripetal relations to the outer world. There are two reasons for this: The patient who wishes to isolate himself from reality must permit the environment to act upon him as little as possible, but he must also not wish to influence it actively himself. For two reasons: By doing so he would become distracted from within and obliged to heed the external world so as to be able to act upon it; furthermore, through the action himself he would create new sensory stimuli and other relations with reality. The autistic and negativistic patients are therefore mostly inactive; [17] they have actively as well as passively narrowed relations with the outer world.

But the autistic patients have, not alone, a positive reason for busying themselves undisturbed with their own ideas where they see their wishes fulfilled. The imagined happiness is not absolute. It is destroyed not only through the influences of the outer world and the conception of reality, but in its place appears much oftener at once under such circumstances, the sensation of the opposite, of the, in reality, unfulfilled wish. All these patients have a life wound, which is split off from the ego as well as may be, and hidden by an opposite conception. For that reason they must defend themselves against any contact with their complex; and, as in the split-up thought process of the schizophrenic, everything, so to say, may have its association to the complex, so everything may be painful to them that comes from the outside. This genesis of negativistic phenomena may often be established through observation or direct experiment when touching of the complexes calls forth the negativism, where it would not otherwise appear.

With this conduct the patients exaggerate and caricature only one of the usual manifestations of the normal. It is a general experience, that questions, which relate to complexes, are at once answered in the negative, even when the persons wish to be open, and afterwards speak of it without dissembling. For there exists an instinctive tendency to conceal the complex. Normal persons, likewise see to it that their life's wound is not touched upon, and they also often have in misfortune the tendency, to withdraw within themselves, because by contact with others there are so many things that root up the pains, by associations with the complex. Even in consequence of bodily pains, which can not be relieved, we often see negativistic conduct, especially in children, who under such circumstances draw back and become repellant in the same manner as our patients, sometimes towards all influences from the outer world, sometimes only under special conditions.

Among children we see still other grounds for negativistic conduct: they often do not understand what is expected of them and turn against the unknown through general obstinacy, for example, during a medical examination, or in being photographed. We observe the same thing in imbeciles, the deaf and dumb and partially deaf, the dream state in epileptics, and in timid or obstinate animals. Schizophrenics also are frequently no longer able to understand the environment, and must become, in the same manner, cross and repelling, although in the course of the disease, the general blind resistance which under normal conditions is to their interest, becomes a detriment.

From the standpoint of the patients the environment moreover frequently appears not only not understandable but directly inimical; at best it does not enter into their needs. We incarcerate them in an institution, rob them of their personal rights; they do not wish to concern themselves about the world, and we wish to force them to; they have ideas of grandeur which are not recognized; they wish to love without being able to command an object; they are persecuted and find no protection, but instead, misunderstanding and refusals.

In the institution the physicians and attendants become the incarnations of such disappointments, while the other patients sympathize with the patient mostly not at all or only superficially, and at any rate stimulate the complexes much less than the officers of the institution. The relatives are sometimes drawn into the complexes, sometimes not. The difference in the effect in the negativism towards different persons is thus easily comprehended. It is just as intelligible that negativism will be called out or increased by opposition from outside, but is dispersed through the greatest possible nonchalance.

The affectivity of schizophrenia contains, furthermore, an additional root of negativism. In the beginning of the disease especially we can often observe a more marked touchiness of the affect, and there is much to indicate the existence of, in the later course, a pathological irritability. Under such circumstances we see, in patients who are not schizophrenics (for example, in neurasthenics) as in schizophrenics who are still capable of social relations, a drawing into themselves, the greatest possible avoidance of all stimuli and a reaction to influences which differs from negativism quantitatively only. Naturally the negativism originating from other sources produces on its side an analogous underlying affective state, so that irritability and negativism together form a vicious circle.

Increased difficulty of action and thought is a further root of negativism which is clear in some cases. There are many different reasons for this, some of which we probably do not yet know. I have not yet been able to establish, as mentioned, a specific motor disturbance. On the contrary there are phenomena resembling brain pressure. True action is moreover impeded by the disturbed associations, most commonly, however, we find in the schizophrenics a peculiar inability to direct their thoughts. "It thinks " in the patients. The flow of thought is automatic, independent of the will; often it is felt as a most painfully fatiguing compulsion; often also the pressure of thought is a matter of indifference as long as he is left to himself. So soon, however, as he is forced by stimulation from without to change the direction of his thoughts, highly distressing feelings arise in both events, which enforce an attitude of repelling.

That the negativistic repelling very often bears the outspoken stamp of the erotic must be due to a root of the negativism being in the sexuality. This is very easily understandable. The sexuality has normally a strong negativistic component; it shows itself clearest in the opposition of the female against the sexual approach, which we find in animals and also in man, where the sexual act is desired. [18] We know that there is no case of schizophrenia in whose complexes sexuality does not play a prominent role, and very often the repelling is founded in sexual delusions, the patients believing themselves loved or violated.

In general, negativism has a close relationship to delusions and hallucinations. These can naturally not lead to a true negativism but to conduct, that can not at all be differentiated directly from negativism, and as delusions and negativism, for the most part, appear side by side, it is wholly impossible, to separate the part played by one factor from that played by the other. The difficulty is increased through the fact that delusions and especially hallucinations are often the sequelae, or better, the expression of negativism; the voices do not necessarily express the negativistic state of feeling but may correspond to another affect. Indeed very frequently the delusion is stated by the patient afterwards falsely as the reason for the negativistic conduct. A young woman, with whom during a paroxysm, one could establish fairly good communication in spite of the negativism, declared afterward, however, that she thought some one tried to hypnotize her and then offer violence to her, on that account she had always done the opposite of what was desired of her.

Up to this point the description has dealt with passive negativism which opposes itself against any demand coming from outside. The resistance leads naturally to active defense, abuse, and to blows, but the doing of the opposite of what is demanded requires a special motivation which in part suggests itself. He who will not open the mouth on request, voluntarily clenches his jaws; he who answers to the request to go forward by walking backwards, is best guarded against a sudden surprise which might insist upon the carrying out of the command; he who will not sit in a certain place indicated sits better in another; he who will not eat his own portion must take another's or go hungry. In short, the opposite action is in most cases so nearby, so self-evident, it so emphasizes the denial and provides such a good position of defense, that it is very apt to be used instinctively by both man and animals.

Not rarely, however, the contrary action so far over-shoots the mark that the hitherto utilized motive for its explanation is no longer sufficient. Thus a patient, who wishes to go to bed and has undressed, receiving a care-less command from the attendant to go to bed, at once begins to put his clothes on again. By error a patient is given cabbage, among other things, on her plate, when it is well known that she does not like it. As she usually gets no dessert, unless she has finished, she is told she need not eat the cabbage; now she eats only the cabbage and leaves the other better liked things on her plate. This same catatonic plays the piano; as soon as she notices that she is listened to with pleasure, she stops; she looks curiously at everything unusual, but at once turns away, however, if anyone pays any attention to her. When she hears an accidental remark: "Now she is doing that," she stops at once, or does the opposite. — Here belong also the forbidden actions. There are patients who will do nothing except what is forbidden them, so that one can make use of this peculiarity. — Or the patient will not carry out an action until it is too late or is no longer pos-sible. So it is quite usual that they first draw back the hand that they should reach out but at once extend it as soon as one turns away from them, or that they give no answer so long as one busies himself with them, but begin to speak when one turns to other patients or when one is about to leave the room. It may also happen that schizophrenics will speak for others but are dumb when asked questions themselves (whether indifferent or important is irrelevant), or when they might have wishes of their own to express.

In these cases, in which the negativism leads to actions, of course those explanations no longer suffice which explain it with the need for rest or the difficulty of the procedure. The inimical relation of the environment could rather be considered as the root for such conduct, but it is absent in many such cases and shows no parallelism with negativism where it is present. Therefore, there must be still other causes of negativism.

The tendency to generalization of single symptoms, always demonstrable in schizophrenia, first suggests itself. Stereotypies, resistances, etc., which are well founded in some occasion, readily expand and become fixed, or at least come to light on many occasions where they are out of place.

A schizophrenic may be imagined as so working up his evasions, that he carries them out when the situation does not demand it and in a manner which is in contradiction with his original (unconscious) object. I do not know how often negativistic symptoms are to be explained in this manner, but when one closely observes the individual patients, one gets the impression that the tendency to generalization does not commonly lead to exactly such conduct.

Ambitendency and ambivalency are of by far greater significance. Both of these two related characteristics, especially the latter, are immeasurably increased in schizophrenia.

I formerly, rather one-sidedly, applied the term negative suggestibility [19] to the psychological fact that a definite tendency to contrary or opposite action is combined with every impulse, whether coming from within or without. I would now prefer to designate the whole idea as "ambitendency." Even in health the negative constituent often gets the upper hand; so soon as one has decided on something, the feeling comes that one had better have done the opposite; people with weak will are therefore prevented from acting. In the territory of the unconscious the opposite impulse often runs counter to our wish. More especially one wishes to be potent on his wedding night, and exactly then most commonly, a transitory impotence occurs. When for any reason the menses are especially awaited, then particularly they fail, etc.

But there are exceptions. As a rule the normal person allows the pro and con to act together, as the physicist works with two forces in opposite directions in such a way that the resultant is governed by the stronger impulse. But, however, as there are always two tendencies, it needs only a small disturbance of their balanced relations, in order to bring out one of them, and this can as well be the negative as the positive one.

In schizophrenia, however, several such disturbances are present. It lies in the character of the disease, that the inter-association of ideas is loosened: each thought, each tendency can exist for itself, without influencing the others and being influenced by the others. Thus a catatonic seats herself at a strange table, cordially assures those standing about: "have no anxiety, I am going to take nothing," serves herself, however, at the same time with sweets and chews with her mouth full. She, or rather, something in her, knows that she should not help herself; that it is disagreeable to those about for her to eat at the table prepared for the guests; she therefore sooths the onlookers, and imagines herself, as not taking anything, but another component of her split psyche longs for the good things and lays to. The two psychisms, which in health would be united in an action of choice, go along here side by side without in the least influencing each other.

While in the above observation the two impulses have become simultaneously active it is also possible for only one impulse to become active at a given moment, giving the other free play to be operative later on. Each goal by itself may dominate the patient for a certain length of time making him the sport of his different impulses. Whether he acts in a positive or negative sense is a matter of accident more or less. Also an already carried out action can be annulled; as when a patient destroys a fully completed piece of work. The negative and positive tendencies can also change very quickly, even during the carrying out of an action. "Not seldom we observe a vacillation in the strength of the positive and negative tendencies; sometimes one, sometimes the other, gains the supremacy. There comes a sudden stand-still and then, just as suddenly, a continuation of the original movement; it continues by fits and starts and becomes angular and awkward." [20]

Kraepelin explains this by the absence of the guiding influence of permanent endeavors and volitional tendencies upon actions. A better expression would be to say that the goal is constantly changing. Gross seeks the pathogeny in the loss of the "highest psychic function." The idea of the latter is very vague. The "synthesis" of the different trends, an expression, which is used by the French for a quite similar conception, is rather a general characteristic of the normal psyche; naturally, like many others, it can become relatively easily disturbed, because it is proportionally complicated. It is not lacking, however, in children, idiots, or animals, only, corresponding to the greater simplicity of such psyches, less developed. It thus becomes difficult to designate this association of different correlated ideas and trends, which suffers first in schizophrenia, in a unity as the highest psychic function. What we observe is just the splitting, the independence of single psychisms, and we will indeed do well, in this obscure territory not to go beyond the observations.

In schizophrenia the stimulus from the outside produces quite as easily, negative and positive reactions: The negative suggestibility is pathologically increased. The building up of negative and positive suggestibility goes along, for the most part, hand in hand. Children, senile dements, and other sorts of affective people are under certain circumstances very easily suggestible; they are, however, quite as often stubborn and negativistic against outer influences. Some authors have long maintained that hysterics suffer from excessive suggestibility, while others deny suggestibility from without; and refer it all to autosuggestion. In reality both peculiarities exist side by side; they are only different sides of one and the same element of character. Certainly, the preponderance of protestations, as already mentioned, has, often besides, the significance of a sort of protection against the exaggerated suggestibility.

In schizophrenia especially, Kraepelin has quite correctly brought negativism into relation with abnormal suggestibility, which expresses itself in command automatism. We often see in the same patient negativism and command automatism side by side, indeed the one may pass into the other. Schizophrenics, like children, swing from one extreme to the other. It must be added that these two characteristics do not always occur together. The relation, even in schizophrenia, is complicated in such a manner as to resist reduction to a simple formula. Schizophrenics, nevertheless, as a whole, in spite of their autistic seclusion from outside, are found to be remarkably suggestible by close examination. Fellow patients who are the ringleaders of a ward find the schizophrenics an easy butt, and for the spiritus loci there is no more delicate reagent for the local color of an institution than the apparently isolated mass of its schizophrenics.

Kleist [21] denies the connection of "negative suggestibility" with negativism. This author has the decided merit of having enlarged upon Wernicke's ideas, of carrying them to their end and presenting them clearly. It is thus a duty to come to an understanding with him. In the first place he cavils at the conception that inhibition should occur in the field of motility as the result of the contrary conception which arises with each idea, constituting a peculiar disturbance in the course of ideation for which brain pathology has no analogy. Here comes out very strongly the difference in methods of investigation. Brain pathology analogies have proved themselves so unfruitful in psychiatry, [22] that to begin with we do not care whether we find them or not. On the contrary we seek analogies in the thinking of the healthy, and then this so characteristic inhibition shows itself to be neither peculiar nor strange. So among the normal many conclusions and actions are stopped in this manner either temporarily or continuously.

Kleist further opposes, that to many ideas there are no contrary ideas, and that a negativistic patient who is requested to pick out the red wools certainly would not choose the green. Here the author confuses the intellectual contrary idea with the affective — the voluntary. We are only considering the latter. Kleist moreover fails to consider that I expressly assume different genetic forms of negativism and designate negative suggestibility as only one of several roots.

Ambivalence. — By ambivalence is to be understood the specific schizophrenic characteristic, to accompany identical ideas or concepts at the same time with positive as well as negative feelings (affective ambivalence), to will and not to will at the same time the identical action? (ambivalence of the will) and to think the same thoughts at once negatively and positively (intellectual ambivalence).

In the case of an idea which arouses both negative and positive feelings the difference is not always sharply appreciated even in health, or otherwise expressed, when a normal person loves something or somebody on account of one quality but hates them on account of another, the result is not an entirely unitary feeling tone, either the positive, or the negative outweighing at times. [23] The ultimate conclusions are not necessarily drawn by the split psyche of the schizophrenic. The mentally sick wife loves her husband on account of his good qualities and hates him at the same time on account of his bad ones, and her attitude towards each side is as though the other did not exist.

Ambivalence of the will or voluntary ambivalence is the natural outcome of affective ambivalence. Intellectual ambivalence needs special consideration. It is of course in close association with affective ambivalence in many judgments but not in all. Even from the purely intellectual point of view each thought is in many ways most closely akin to its opposite; not only that the closest association to "white" is "black": each judgment contains the negation of its opposite, and there would be no sense in thinking it unless the contrary had entered into consideration: I can not think and say: "the sky is blue," unless the contrary, that it may not be blue, is, so to say, in the air. [24] Censure of a picture lies psychologically much nearer praise of the picture than any other thought. Children frequently use the same expressions for both positive and negative ideas, for example, tü tu for Türe zu (door to) for open and close the door, also "zuletzt" (last) for "zuerst" (first), and later, when they first begin purely in play to judge, they often do not care at all how they express the same. [25]

With the confused schizophrenics the distinction is often completely blotted out. Affective motives also probably cooperate, as in the above mentioned patient, who at the same time censured and praised her husband; but it is probably a purely intellectual fault when a catatonic who after having answered his wife's friendly letter, with an unmotived farewell letter, said, in answer to expostulations: "I could have just as well written another letter, good day or farewell are just the same " (dire bonjour ou dire adieu). So thesis and antithesis in our patients often become so similar as to become confused or even identified one with another.

Ambitendency and ambivalency in themselves bring about only an equalization of correct thoughts and conflicts with their opposites. In negativism, however, these opposites actually gain the ascendant. There are two known reasons for this: In the first place this predilection is certainly often merely apparent. Even the negativistic produces correct thoughts and actions. When, however, among a thousand psychisms in our day only a single one is negativistic it is conspicuous, the probability would be that in the equalization of tendency and antitendency there would be five hundred false to five hundred correct reactions, a proportion which would imply severe negativistic anomalies.

Furthermore the previously mentioned "contradictions with reality," especially autism, take care that the contrary action is favored as much as possible.

Outer negativism is therefore, in the first place, due to a number of factors, which place the patient in opposition to the outer world; the effect of this contrariness can become so extensive because the schizophrenic ambitendency and ambivalency furnish a good soil for it, and above all, remove what in the normal opposes perverse actions.

Ambitendency and ambivalency make inner negativism also somewhat comprehensible to us in some degree, which would not be explainable through other factors which cause negation. When, as in will-negativism, each impulse is opposed by a contrary [26] impulse, and when the psyche is so split that each of these two tendencies can independently assert itself so that a compromise between them is impossible or is made very difficult, then the antitendency will often manifest itself instead of the tendency. It has not been positively demonstrated that it is just this antitendency which asserts itself with especial frequency. It is, however, probable that such cases occur. They would in some degree be intelligible through the inner disruption in which such patients find themselves. They are not pleased with anything, nothing gives them any satisfaction, so it is comprehensible that they seek something else; and that "something else" is very often the opposite.

I believe, however, that there exists besides an unknown factor which gives a special weight to the contrary tendency, not only because the observation of negativistic schizophrenics sometimes appears to point that way, but also because auto-suggestions in the normal are so frequently negative; the menstrual period arrives when it is certainly not expected and vice versa. This factor requires further study.

The cross impulses have very different significance. A part of them are, of course, negativistic. One will not do the desired and so in some cases does the opposite; in others only something else. The apraxiform approximate acts often have the character of acts in emotional confusion under which circumstances the normal make all sorts of errors. Most commonly the cross impulses probably are the result of the specific schizophrenic train of thought in which all at once the nearby association becomes the principle thing; the thought is at once cut off and there is a new one of unknown genesis or at least of insufficient connection with the preceding; or suddenly a quite abrupt thought, an hallucination, an automatic impulse to movement, suddenly arises out of unconscious complexes. It is sufficient only to hint at these things which are self-evident to one who knows dementia praecox.

Intellectual negativism resembles volitional negativism very much. When an idea stimulates its opposite and the thought becomes split and unclear, so that criticism is difficult, the antithesis is apt to acquire undue weight, and under certain circumstances replace the thesis. The latter especially because the patients, with their changed feeling and thinking, are often actually compelled to see the thing in an unusual way. Nevertheless, cases like the one previously mentioned, in which each thought compelled the thinking of a contrary thought, give cause for the conjecture, that here preference escapes us as a factor that leads to the contrary thought. Also the dream of the normal, in which many an idea is represented by its opposite, appears to me to point to an active predilection for the negative. Perhaps also the mechanism of wit, which often replaces one thing by its opposite, has a point of contact with intellectual negativism.

It has also occurred to us that inner negativism, especially the intellectual, might express itself in experiments in negative or contrast associations. This conjecture has not been established by proof; we have only seen a striking tendency to contrast association in two patients, and precisely these were not negativistic.

R. Voght [27] on the basis of the views of Lipps, propounds a hypothesis which might explain, in hysterics, how an idea may inhibit precisely the closely related and therefore other to-be anticipated concepts. I believe the supposed identification of transfer of energy and association therein set forth is too visionary to warrant discussion.

Inner negativism is much rarer in both its forms than outer negativism. This is easily understood after we have seen how much outer negativism is favored by the disturbed relations to the environment, which are constantly present, but favor will-negativism only slightly, and intellectual negativism even less. Negativistic phenomena can not so easily originate, of course, exclusively upon the basis of ambitendency and ambivalency, the predisposing factors of negativism.

Notes

[1] E. Bleuler, Zur Theorie des schizophrenen Negativismus, Psychiatrisch-Neurologische Wochenschrift, Vol. 12, 1910/11, Nr. 18, 19, 20, 21.

[2] Psychiatric, 8. Aufl., I, 380. Barth. Leipzig, 1909.

[3] In: Binswanger und Siemerling, Lehrbuch der Psychiatrie. Fischer, Jena, 1904. S. 258.

[4] Beitrag zur klin. Analyse des Negativisums bei Geisteskranken. Zentralblatt f. Nervenkrankh. und Psychiatrie, 1902, S. 554.

[5] Zur Psychophysiologic des Negativisums, Zentralblatt f. Nervenheillc u. Psychiatrie, 1903, S. 85.

[6] Motorische Störungen bei einfachem Irresein, Allgem. Zeitschrift f. Psychiatrie, Bd. 42, S. i.

[7] Zur Genese einiger Symptome in katatonen Zustanden. Neurol. Centralbl., 1904, S. 8.

[8] Psychiatric, Aufl. I, S. 453.

[9] Psych. Kontrasterscheinungen bei einer Geisteskranken, Arch. ital. per le malatie nerv., Bd. 24, Ref.; Allgem. Zeitschrift f. Psychiatric, 1887, S. 58.

[10] Negativismo vesanico e allucinazioni antagonistici, Bull, della soc. Lancisiana degli osped. di Roma, XVI, 96. Ref. Zeitschrift f. Psych, u. Physiol, der Sinnesorgan, Bd. 13, S. 397.

[11] Centralbl. f. Nervenhkd. u. Ps., 1906, 622 (Dromard).

[12] Nerven- und Geisteserkrankungen in der Zeit der Geschlechtsreife, Wiener klin. Wochenschrift, 1904, S. 1 161.

[13] A. Zeitschrift, Bd. 58, S. 226.

[14] Die Affektlage der Ablehunng. Monatsschrift für Psychiatric und Neurologic, 1902, Bd. XII, S. 359. Beitrag zur Pathologie des Negativisums. Psychiatrisch-neurol. Wochenschrift, 1903, V. Jahrg., S. 269. — Zur Differential-Diagnostik negativistischer Phänomene. Psychiatr.-Neurol. Wochenschrift, 1904, Bd. VI, S. 345.

[15] Weitere Untersuchungen von Geisteskranken mit psychomotorischen Storüngen. Leipzig, Klinkhardt, 1909, S. 97 f.

[16] By autistic I understand practically what Freud (not however Havelock Ellis) means by autoerotism. I think it well, however, to avoid the latter expression, as it is misunderstood by all those not very familiar with Freud's writing. I have discussed this at length in the chapter on Schizophrenia in Aschaffenburg's Hand-Book of Psychiatry. The symptom of ambivalence to be mentioned later in the text is also discussed in this book.

[17] Under special circumstances this seclusion may be overcome as in the acute hyperkinesias, in which the movements result from an impulse, and in paranoids, who, while the autism is not fully complete, are sensible of the interference with their wishes, and translate them into delusions of persecution and react accordingly. In both cases there is a much-narrowed relation to the outer world.

[18] This opposition lies apparently at the bottom of the aversion of the cultured to think or speak of sexual things. I certainly do not underrate the role of artificial convenience; this convenience, however, which leads to so much disadvantage and nonsense, must be grounded in our nature, otherwise it would not have developed.

[19] Psych.-Neurol. Wochenschrift, 1904, VI Bd., Nr. 27/28.

[20] Kraepelin, Psychiatrie, Achte Auflage, I, 373.

[21] Kleist, Weitere Untersuchungen an Geisteskranken mit psychomotorischen Storungen. Leipzig, Klinkhardt, 1909, S. 97 f.

[22] Cerebral pathology and localization ideas have led so extraordinarily capable an observer and fruitful thinker as Wernicke into sterile by-ways.

[23] A normal ambivalent group of ideas is represented by the sexual, especially in women, as stated previously.

[24] "Each idea demands, as it were, a contrary idea as its natural complement." Wreschner, Reproduktion und Assoziation von Vorstellungen, Zeitschr. für Psych, u. Phys. der. Sinnes-O., Ergänzungsband 3, 07/9, S. 595-

[25] Compare also the latin "religio," that was used in both a good and a bad sense, as a blessing and as a curse. Also see van Ginnecken (Principe de linguistique psychologique, Leipzig, 1907), who goes rather too far.

[26] The contrary impulse often consists in the not carrying out of the original intention. From our view point to do something and not to do it is a contrast, just as to do something and to do the opposite.

[27] Die hyst. Dissoziat. im Lichte der Lehre von der Energie-Absorption. Zentralbl. f. Nervenheilk. u. Psychiatrie, 1906, S. 249.

www.ingramcontent.com/pod-product-compliance
Lightning Source LLC
Chambersburg PA
CBHW021049180526
45163CB00005B/2356